WHY IS MY CAT DOING THAT?

Sarah Heath

UNDERSTAND YOUR PET'S PUZZLING BEHAVIOUR

hamlyn

To Matt and David – for being the most
important people in my life and teaching
me what really matters.

An Hachette UK Company
www.hachette.co.uk

First published in Great Britain in 2009 by
Hamlyn, a division of Octopus Publishing Group Ltd
2–4 Heron Quays, London E14 4JP
www.octopusbooks.co.uk

ISBN 978-0-600-62023-5

A CIP catalogue record for this book is
available from the British Library.

Printed and bound in China

2 4 6 8 10 9 7 5 3 1

The advice in this book is provided
as general information only. It is not necessarily
specific to any individual case and is not a
substitute for the guidance and advice provided
by a licensed veterinary practitioner consulted in
any particular situation. Octopus Publishing Group
accepts no liability or responsibility for any
consequences arising from the use of or reliance upon
the information contained herein. No cats or kittens
were harmed in the making of this book.

Unless the information in this book is specifically for
female cats, cats are referred to throughout as 'he'.
Unless otherwise specified the information is
applicable equally to male and female cats.

Contents

Introduction

Cats have grown increasingly popular in recent years and have now surpassed dogs as the most popular pet in many western countries. They provide an attractive combination of loving companionship and independence that meets the needs of busy pet owners. Yet cats remain enigmatic to us in many ways and some of their behaviours leave us longing to understand more about them. This book gives an insight into the natural behaviour of this fascinating species and helps to take the mystery out of some of their actions.

Although cats share our homes and lives, they are not a naturally pack-living species and their relationship with us is one of companionship rather than dependence. Feline survival is ultimately a solitary affair and social interactions are based on a desire to be together rather than a need for social support. Owning a cat is therefore something of a privilege and we can feel flattered when they choose to spend time with us. Appreciation of the social behaviour of cats helps to explain how apparent devotion and indignant independence can co-exist so readily, making sense of why our cats can snub us one minute and curl up on our laps the next.

Cats communicate in different ways from us and their signals can be difficult to interpret with human eyes. Feline body language can be very subtle and to an untrained eye much of what they say to each other, and to us, goes unnoticed.

Pouncing on prey is a natural feline behaviour.

While bathing in the sun this cat shows off some of its impressive weaponry.

When feline signals are more extreme, misinterpretation threatens to colour our judgement and cats that show aggressive signals are inappropriately labelled as malicious.

Scent communication is very important to cats but the human sense of smell is poor and we simply lack the olfactory equipment to decipher the messages that scent signals contain. Sometimes this results in miscommunication that threatens the pet–owner relationship and tolerance of marking behaviour when it occurs indoors is certainly very low.

We love to watch our cats interact with each other. The decline in tactile contact as they mature from cute kittens into independent adults can be mistaken for a breakdown in their relationship and leave us worrying that our pets feel lonely. A better understanding of feline behaviour prevents us from falling into this trap and helps us to appreciate the quality of their less dependent adult relationships. In fact, too many cats live their lives in close proximity to feline strangers and are forced to share their essential resources with cats that they do not like.

By respecting their solitary survivor status and limiting the numbers of cats we keep to socially compatible levels, we can save them from the endurance of chronic stress with all the behavioural and medical challenges that it brings.

While it is understandable that some owners find the hunting activity of their cats distasteful, we can still admire the skill of this behaviour. By appreciating the drives behind the hunting instinct we can also improve the quality of play that we offer to our cats and channel some of their prowess into more acceptable outlets.

Through a series of photographs this book seeks to unravel the mystery of feline behaviour, improve your understanding of how your feline friends think and increase the value of your relationships with them. Understanding some of the perplexing behaviours that our cats display can add a new dimension to our interactions with them and decrease our perception of some of their activities as problems. Cats are fascinating, adaptable and independent creatures that can teach us a great deal about companionship and reward us with unconditional affection. In return, we seek to understand them better and use that knowledge to give them the best quality of life we can.

INDEPENDENT SPIRITS

The family

The way that cats interact is driven by their natural instincts. As adults they are socially independent but when they are young kittens they have a greater need for social contact, with their mother and their littermates. They learn important lessons from interactions with each other in their first weeks.

Cuddle up!

Spending time with littermates is very important for kittens. At this age they still have a need for social contact and they will spend more time in physical contact with each other than adult cats. Lying close to each other gives them confidence and also creates the opportunity to take some cute photos!

Safety in numbers

The presence of the mother also offers emotional support and gives kittens confidence. These kittens sit close to their mother and to each other and from this position of safety they can show interest in the approaching person. Their ultimate response will be influenced by how the queen reacts. If she is hostile they will learn that people are best avoided but if she is friendly they will be keen to say hello.

Family first

The first important relationship in a kitten's life is that with his mum. These kittens are entirely dependent on the queen for survival and she takes care of all of their physical needs. She provides nutrition through suckling and deals with her kittens' toileting needs by licking their abdomen to stimulate the passing of urine and faeces that she can then consume.

Kitten love

Relationships within a litter and between a mother and her kittens are particularly strong because the kittens are dependent on social contact for survival at this stage of their development. Cats have been referred to as the first true feminists and the matriarchal nature of feline society is central to their social behaviour. Sometimes this can make their behaviour difficult to fathom from a human perspective!

Kitty crèche

Cats live in a matriarchal society and related females work together to rear their young. Each queen has an interest in the survival of all the kittens, as they share genetic material with one another. There is safety in numbers and when the kittens are very young queens may suckle the offspring of their female relatives – this helps to give the kittens more extensive antibody protection from disease.

I wonder what's over there?

As kittens get older they start to develop more independent behaviours and will begin to venture away from the nest area. At two weeks of age these kittens are becoming more mobile but they are still dependent on their mother who is popping back to the nest to check that her offspring are clean.

This is fun!

The play between kittens is much more than a way to pass the time. It offers the opportunity to learn how to communicate effectively with other cats and also to develop important adult behaviours such as predatory skills. Single kittens are known to be at risk of developing behaviour problems later in life because they are denied the chance to spend this valuable time with kittens of the same age.

Social or anti-social

Feline social behaviour can sometimes seem very perplexing. Cats certainly do seem to develop meaningful friendships with some cats but at the same time can seem aloof and disinterested in others. Behavioural signs can help to identify those cats that are considered part of the same social group and communication through touch is particularly important in this context. Being alone is not a negative experience for cats since their wild ancestors did not depend on each other for survival.

In good company

Members of this group of related individuals naturally spend time in close proximity to one another. Some of them are actively sniffing at the ground, indicating that there is important information about other cats that may have passed this way. Their relaxed body postures show that they are very comfortable in each others' presence.

Scratch my back

In adulthood cats do develop special relationships with each other. These two cats demonstrate their friendship through tactile communication. They are lying touching one another and will rub against each other and groom each other as a way of sharing scent. These behaviours confirm their friendship and ensure that both of the cats feel safe in each other's presence.

Mystery man

The adult tom cat is an independent creature. While he is capable of forming social relationships with other cats his survival is ultimately his own responsibility. His aim is to avoid conflict with other cats that might leave him injured and vulnerable, therefore he spends a lot of time alone.

Keep your distance

The supposedly characteristic arrogance and aloofness of the cat is often supported by images of cats on their own or of awkward 'stand-off' encounters. When cats turn their backs on each other or elect to spend time alone, their owners may interpret them as being snooty and self-sufficient but often their attempts to limit social contact are actually driven by a lack of confidence and a desire to avoid confrontation.

I'm not interested

Stand-offs are common in the cat world. These two individuals are well aware of each other's presence and the position of the tabby cat's ears indicate that he is monitoring the other cat. Neither wants to engage in any form of social interaction and by staying still they are less likely to attract the interest of the other.

Stalemate

When cats find themselves in close proximity to other individuals their body posture often displays a degree of tension. In this case it is the ear position of the tabby in the foreground that indicates that he is keeping tabs on the activity of the ginger cat behind him. Both cats are showing inhibited behaviour and will be reluctant to move away from their positions for fear of inciting interest and possible conflict from the other cat.

King of the castle

This cat has selected an elevated resting place which is only big enough for one! From this safe position he can afford to bluff about his ability to defend himself and his arched back and fluffed-up tail are designed to make him look bigger than he really is.

Food and drink

Providing food for those that we care about is a very important nurturing behaviour for humans and eating is a key component of our social behaviour. We invite our friends round for dinner or serve them with tea and biscuits when they call round on a Sunday afternoon. So, when cats show little interest in their food or seem less than impressed by the treats we offer, it can be difficult for us to understand.

Too close for comfort

These cats all live together and their owner provides them with one bowl of food to share. Although all the cats have come to eat, their body language shows that they would much rather eat alone. The cat on the left shows an extended body posture with its back legs poised for running away if the tension gets too much. The other two cats are sitting down to eat but their ear posture shows that they are not relaxed.

Stay away

In the cat world eating is purely a refuelling exercise and does not have any social significance. Cats would therefore prefer to eat alone as this cat is doing and they can become very tense if they are disturbed when they are eating. This cat is showing his disapproval of the approach of another cat by staring at it as he continues to eat.

Lapping it up

Cats prefer to drink water from containers that are ceramic, metal or glass, as plastic can taint the water and make them wary of drinking. This picture shows how keeping the water filled up to the brim enables the cat to lap from the surface without its whiskers touching the side of the bowl.

Saying hello

Owners of more than one cat often find it hard to tell if their cats are friends. They spend a great deal of time telling each other to keep their distance but they also use body language and vocalization to signal when they wish to engage in contact. Cats maintain friendships through the use of affiliative behaviours, such as allogrooming and allorubbing, which they also use in interactions with humans or other species they accept as friends.

Rub me up the right way

Rubbing on one another (allorubbing) is an important behaviour between cats that enjoy a close relationship. These related cats are showing a typical rubbing interaction, which starts with the head rub and then often continues with the cats rubbing their body length against each other and even entwining their tails.

Tail-tip greeting

The tail-up posture seen in this picture is the typical feline greeting. When a cat walks toward another cat or person with the tail held high it is a clear indication that he wishes to say hello. The tail tip is usually curved over slightly, as it is here, and the way in which the tip curves differs considerably between individuals.

You belong to me

When cats enjoy a close relationship with
their owners they will rub against
their legs in an act of allorubbing.
This cat is saying hello to its owner
and exchanging its scent with theirs.
This helps to identify the person as part
of the cat's social circle.

Hold still while I wash you

Grooming each other (allogrooming) is another way in
which cats cement their relationship. Usually one cat
will initiate the grooming, then the other will reciprocate.
These two related cats are enjoying each other's company
and one cat uses his front legs to hold onto the other while
he washes her ear. There is no resistance from the other
cat and she is likely to return the favour when her friend
has finished.

Dogs are from Mars, cats are from Venus

Differences in body language between dogs and cats can sometimes lead to misunderstandings. This young Pug is interpreting the cat's behaviour as a sign of play but the cat's sideways body position, the arched back and the fluffed-up tail are signs of fear. The way the cat has distributed his weight is a sign that he is preparing to swipe out as he runs away.

We're family

Allorubbing can be used to reinforce any social relationship, including those with other species such as humans and dogs. This cat and dog live together and know each other well. The act of the cat rubbing his face along the side of the dog's face is an indication of the affectionate relationship between them.

How do you smell?

These cats are greeting each other with nose-to-nose contact that conveys important information about one another. This sort of relaxed greeting involves the close proximity that is normally only possible if the cats are already familiar with one another.

Pay me attention

When cats roll onto their sides the action is referred to as a 'social roll'. It means that the cat wants to initiate a social interaction and cats will often perform this behaviour after a period of separation from their feline friend or owner. The invitation is for passive social contact so it is important not to be fooled into thinking that the cat wants you to touch its belly!

Cats and humans

We derive an enormous amount of enjoyment from our relationships with our pet cats but in order to ensure that they also receive the greatest benefits from living with us, we need to teach small kittens appropriate lessons about human handling. Misunderstandings between the species can result in confusion for us and a significant level of stress for the cat.

Don't rush in

Kittens and children can make great playmates but it is important to look at the world from a cat's perspective to make sure that interactions between them are beneficial to both. This kitten is unsure about the approach of the child. His weight distribution suggests that any rapid advance from the little boy could be met with a swipe of his paw and potential injury for the child.

Gently does it

Kittens need to experience regular human handling from as early as two weeks of age and the style of handling is important to prepare them for living with people. They need to learn that it is not threatening to be lifted off the ground and it is advisable to use minimal restraint to ensure that the experience is as positive as possible.

I'm just happy!

Adult humans can also misinterpret cat communication and this body-rubbing greeting is often mistakenly seen as a form of begging. When owners respond by feeding the cat they initiate a learning sequence that can rapidly lead to overfeeding and even problems of obesity.

Put me down

This kitten is not having a positive experience and is likely to learn that human handling is something to avoid. Turning a cat onto his back makes him feel vulnerable and this kitten is preparing to lash out with both of his front paws in an act of defence. Handling kittens in this way creates negative associations with being picked up and can even induce aggression toward people later in life.

Catty interactions

The ways in which cats communicate with each other are sometimes difficult for humans to understand and it can be particularly challenging to find out what cats think about each other! Certain types of behaviour can reveal feline friendships, while understanding the importance of rubbing and grooming interactions from a feline perspective can help us to appreciate our pet's view of the other cats with whom they share their world.

I like you

Kittens are far more likely to spend time in close proximity to each other than adult cats but when cats are related or share a close relationship they will continue to cuddle up to each other. The darker ginger cat on the right is initiating a session of mutual rubbing and the other cat is passively accepting.

You missed a spot

Grooming is a very important feline behaviour and cats will spend a great deal of their time grooming themselves. As kittens they accept grooming interaction from their mothers and in adulthood this acceptance of mutual grooming is reserved for members of the same social group. This tabby cat looks less than impressed by the behaviour of his black and white friend but the fact that he is accepting this intense grooming tells us that they know each other well.

I'm not looking for a fight

When unfamiliar adult cats find themselves in close proximity their natural reaction is to slow down and approach with caution – rapid movement could induce chase and potential injury. These cats are exchanging a combination of visual, vocal and olfactory communication signals that are all designed to ensure that physical confrontation is avoided.

CATS' EYES AND MORE

Hunting skills

Cats have a highly developed sensory system that not only helps them to locate their prey but also gives them an elaborate system of communication using the senses of sight, sound and smell. Cats do not live in packs as adults and their survival is very much a solitary business, so they rely heavily on signals that enable them to communicate with each other from a distance.

I'm stalking you

This cat is using slow stealth to get closer to its intended prey and the chickens seem to be blissfully unaware of its presence. Cats do not hunt in packs and it can be a challenge to bring down prey of substantial size, so the chickens may be confident that they are not in real danger.

What do I do now?

This young cat has the characteristically alert ears, eyes and whiskers of a cat on the hunt but appears to be bemused by its potential catch! His front leg is extended as he paws at his victim but in common with many naïve cats, this youngster hesitates instead of continuing with a swift bite at the neck (nape bite) to dispatch his prey.

The huntress

Cats use a combination of sound and visual information to locate their prey. This cat has adopted a lowered body posture and is moving slowly toward its intended target. The forward pricked ears and fixed stare are characteristic of a hunting feline but the nose and the whiskers are also playing their part in ensuring accurate location of the potential kill.

Mixed messages

The use of urine and faeces in communication is something that we do not come across in human society and thus find particularly hard to understand! As a result urine spraying and middening (the depositing of faeces as a communication signal) are behaviours that are often referred to as problems but when they are performed outdoors they are a perfectly normal part of everyday feline language. Scratching is another natural behaviour that can be difficult for us to comprehend.

SCENT MESSAGES

Cats have specialized organs which enable them to process scent messages which can modify their emotional state and their behaviour. Pheromone products can be very beneficial when treating cats that are anxious.

Smelly sign posts

The spraying posture ensures that urine messages are deposited at nose height and can easily be read by passing cats. This kitten sniffs intensely at the message but it is not a threat. It is just a way of transferring information from one cat to another across time.

Scratch and sniff

With the body at full stretch, this cat can get sufficient purchase on the surface to exercise its muscles and tendons and to remove the blunted outer sheath of its front claws. In addition to this functional side of scratching there is a communication element to this behaviour. Scent glands on the pads of the feet are expressed onto the surface of the wood as the cat scratches, to leave an olfactory signal to other cats that pass through the territory.

My territory

This cat adopts a urine spraying posture to leave a message to other cats passing through the territory. He is listening for activity around him and his mark is intended to ensure that he does not have unwanted encounters with feline strangers.

Don't invade my space

The use of faeces as a communication tool is less common but cats do sometimes defecate in the middle of a lawn without making any attempt to cover their deposit. It is being left as a clear scent and visual signal that the territory is occupied and that other cats should stay away.

Talking cats

Vocalization is a far more familiar communication tool for people and we rely heavily on conversation to get our message across. Cats also use their voices but the range of meows, purrs and hisses can sometimes be difficult for us to interpret accurately. Learning more about feline vocal communication will help you to understand what your cat is saying when it speaks to you and thereby improve your relationship.

Keep back

Vocal signals can be used to indicate a negative emotional state and the sounds that are produced while a cat's facial muscles are in high tension are sometimes referred to as strained intensity calls. This cat is confirming its defensive body language with a harsh and penetrating hiss, which is designed to make the threat retreat and thereby reduce the risk of potentially dangerous physical confrontation.

How do you do?

Vocal signals are also used by cats to facilitate social interaction and the main greeting vocalization is the meow. It is used to initiate friendly communication with other cats and with humans. This cat shows a relaxed body posture, with softened whiskers and alert ears, while he looks intently at his owner and utters a gentle meow.

I'm relaxed

Not all vocal communication in the feline world is made by moving the mouth and the classic sign of feline contentment, the purr, is made while the mouth is closed. The purr is an important part of communication between the queen and her litter and is often, though not exclusively, associated with situations of comfort and safety. This cat snuggles down with its owner and responds to the gentle handling with a contented purr.

FELINE AFFECTION

Be my mate

Cats enjoy spending time alone and have a range of signals that enables them to avoid unwelcome encounters but there are times when close proximity is necessary and the breeding season is one of them! Clear communication is needed to ensure that entire cats (cats that have not been neutered) get together so that they can mate and a combination of scent and sound is used to achieve that goal.

Looking for love

Urine spraying is performed more frequently by male and female cats during the breeding season. The urine contains important information about the sexual status of the cats and their readiness to mate. This tom is spraying his mark on vegetation in his territory with the aim of attracting any entire females in the area.

It's all in the timing

Cat mating is a noisy affair and there is a good biological reason for this. In order to maximize the chance of a successful mating, female cats do not ovulate until mating has taken place. This ensures that the egg is released when sperm is available to fertilize it. The ovulation is triggered by barb-like structures on the tom cat's penis, which cause irritation to the female's vaginal wall. This queen is reacting to the sensation with a characteristic scream.

Come and get me!

When female cats are in oestrus they produce scents that indicate their willingness to mate. In order to spread that scent this female is rolling back and forth on the floor and will waft her legs in the air. This un-ladylike behaviour is often displayed on pavements or even in the gutter!

A mother's love

When kittens are very young their relationship with their mother is the most important and she takes care of all of their physical needs. In order to make a rewarding pet in later life, kittens need to transfer their affection to humans and we take on the role of nurturing them. However, there are some aspects of the nurturing behaviour of the queen, especially in the bathroom department, that are more than a little bizarre from a human point of view!

Keeping clean

Very young kittens cannot toilet without being stimulated by their mother licking at their abdomen, as this queen is doing. She consumes the urine and faeces as it is produced, keeping the kitten clean and reducing the risk of elimination deposits transferring infection within the nest area.

KITTEN LESSONS

From birth young kittens gain confidence from the presence of their siblings and choose to spend time in tactile contact with them. Kittens learn important lessons about life through playing with individuals of the same age. When kittens are born as singles and are denied the opportunity for social play, this can adversely affect their development and may predispose them to experience behavioural problems as adult cats.

Human bonding

Feline company is obviously important for kittens but when they are destined to be pets it is also important to incorporate human interaction into their daily routine. Adult felines do not have a high requirement for social contact but owners like to be able to pick their pets up and cuddle them. It is therefore important for this kitten to learn that being lifted off the floor and gently restrained is a positive aspect of living with people.

Mum's reassuring presence

As the kittens grow older they will become less and less physically dependent on their mother. They still benefit from the comfort of her presence, however, and, even though these kittens are old enough to be weaned and eat solid food, they still enjoy snuggling up to their mother and taking a little top up of nutrition by suckling on her.

EMOTIONAL RESPONSES

Scaredy cat

The way in which a cat behaves is driven by the way it feels and understanding feline emotions is the key to interpreting the things they do. Cats cannot talk to us about how they feel and therefore we need to learn how to interpret the communication signals that they use, such as facial expressions, body postures and vocalizations. Cats also use scent to communicate with each other but humans are very poorly equipped to recognize these signals.

I'm going to swipe you

Cats are not very good at getting themselves out of dangerous encounters, therefore they invest a lot of energy in avoiding confrontation in the first place and keeping a safe distance from potential sources of danger. This young kitten has already perfected the art of hissing at an advancing threat but if the threat continues to advance he is prepared for the extra tactic of swiping out with his front paw.

I'm big, honestly!

This tabby kitten has been startled by the appearance of an unfamiliar cat and responds by making himself look more capable than he really is. The arched back and raised tail with the extended back legs, give an impression of increased physical size but the flattened ear posture and forward extended whiskers give more honest information about its state of fear.

I'm *really* scared now

When the threat is even closer cats are more honest about their fear and they adopt a crouched posture, as seen here. The flattened ears, extended whiskers and hissing all signal that this ginger cat is feeling very scared. His weight on his front paws is an indication of his intention to spring backwards and run away if the threat advances.

I'm scared, keep back

Raised hairs over the back and the tail give this young grey cat the impression of increased size. However, the pupils are dilated, creating a staring expression and the weight distribution on the paws shows his readiness to run away if the challenge intensifies.

I feel threatened

When cats are unable to retreat they display more intense body-language signals. Ears are vital for a cat's survival as they detect potential danger and prey. This cat has flattened his ears so they are almost invisible. This posture protects ears from injury during a fight.

Who goes there?

Facial expressions can change more rapidly than body postures and it is the face that gives the most up-to-date information about the cat's emotional state. This cat has been relaxing as indicated by his body posture. His rest has just been interrupted by the approach of a person, however, and the facial expression, with pricked, forward-facing ears, is one of an alert individual that is cautiously interested.

Out of the way

The risk of injury is something that cats work hard to avoid and much of their behaviour is aimed at avoiding confrontation rather than dealing with it. This cat is responding to a situation of potential danger by retreating to a safe haven and physically hiding from the threat. The provision of suitable hiding places is an important part of caring for a nervous cat.

Don't look

These two cats are close to each other but their body language tells us that they are not relaxed. The cat in the foreground has pressure on his front paw pads and is ready to spring away. His ears are swivelled to take in any sound that will enable him to monitor the movement of this adversary. The other cat is also tense and is preparing to move slowly away so as not to induce a chase response. The pricked ears are indicators of his fearful state.

Frustration

Frustration in cats arises when they develop expectations that cannot be met and as a result they can show behaviours that owners find irritating or troublesome. In most cases owners are unaware that their cat is frustrated and interaction with a highly aroused feline in this state can lead to a range of unexpected behavioural responses. Reducing the problematic behaviour depends on changing the underlying emotional state, or more importantly preventing it from developing in the first place.

I want some more

During weaning the queen prepares her kittens for dealing with frustration. She achieves this by terminating suckling sessions while the kittens still have a high expectation of receiving milk. One potential problem with hand-rearing kittens is that humans are more likely to allow the kittens to terminate the suckling session and only remove the bottle when they cease to be interested. This can lead to a range of frustration-related behaviour problems later in life.

I want to get you

This cat is intently watching the activity of another cat out in the garden. The forward pricked ears and slow thrashing tail indicate a high state of expectation and arousal. Access to the other cat is denied by the patio doors and could lead to frustration. Touching this cat while he is in this emotional state may result in him lashing out, as the tension that he feels towards the other cat could be redirected towards the person.

On a mission

This cat is confident and alert. Preventing this individual from doing what he has set out to do could trigger an emotional state of frustration. Individuals that are genetically confident and have learned that they can repeatedly achieve their goals are more prone to frustration and may demonstrate this emotion through behaviours such as pacing and meowing in a demanding manner.

The predator

Hunting is a difficult behaviour for many cat owners to come to terms with. Human love of birds and small rodents can make it hard to accept that a loving family pet can turn into a heartless murderer. However, watching a feline in the hunting sequence highlights the skill of the behaviour and a better understanding of the body language can help us to admire the cat for the successful hunter that it is.

It's mine

Once the prey is caught this cat feels vulnerable and is startled by the presence of a human. The raised hairs over the ridge of the back show that he is feeling fearful and anxious as this cat prepares to hold on tight to his valuable kill. This posture is likely to be accompanied by a low growling vocalization.

Ready to pounce

During the hunting process cats are alert and focused. This cat's forward-facing ears and penetrating visual contact with his victim demonstrate the intensity of his concentration. Meanwhile his body posture and forward weight distribution indicates his readiness to pounce and complete the kill.

Gotcha!

This young cat is focused on his potential prey and has the characteristic erect ears, forward-facing whiskers and fixated stare. However, he betrays his juvenile and inexperienced enthusiasm by entering into the pounce too early in the kill sequence and runs the risk of alerting the prey to his presence and giving it the opportunity to escape!

Relaxation

Relaxation in cats is associated with a range of very appealing behaviours that make this fascinating species even more endearing. Research has shown that watching a still, peaceful cat or stroking his fur also lowers blood pressure in humans, so the relaxation can be infectious. However, some of the responses, such as the social roll, can easily be misinterpreted and can lead to the onset of behaviour that owners may see as problematic.

Catching a few rays

In a secluded corner this cat takes advantage of the sunshine. The posture is relaxed and the paws are not in direct contact with the bench. The tail lies loosely tucked next to the body and the closed eyes and relaxed ears suggest that this individual feels safe enough to take a little nap.

Look, but don't touch

This cat is fully relaxed and shows an alert and interested facial expression while taking all four feet off the ground and exposing his undercarriage. This social roll indicates that he desires social interaction but it is *not* an invitation to stroke his belly. Misinterpreting this behaviour can lead to relaxed cats such as this one being labelled as Jekyll-and Hyde-characters as this apparently friendly invitation is replaced with a seemingly aggressive attack. In fact, clasping the human's hand with all four legs while raking with the back claws is a defensive response to what the cat interprets as a threatening advance from the person.

Just resting

As this cat relaxes in the sunshine he curls up his front paws so that the pads are no longer in contact with the table. This signals that he feels safe and has no intention of running off anywhere. His facial expression indicates that he is not entirely relaxed, however; his slightly rotated ears suggest that he is listening to the activity of others in the garden.

TOILETING BEHAVIOUR

Eliminations

Cats are renowned for their cleanliness and this is often cited as one of the most positive aspects of owning a cat. The natural drive to bury faeces and urine leads to an instinctive digging response to soft, rakeable litter material. As long as kittens have access to suitable litter dirt from an early age, the house-training process is usually rapid and successful. However, when things go wrong and deposits are found in unacceptable locations human tolerance is often very low.

Wrong place

After depositing faeces on the kitchen floor this cat shows a raking response, suggesting that he is trying to bury his deposit. Obviously this is not going to be successful but the raking indicates that this is a case of elimination in an inappropriate place rather than an extreme marking behaviour. In order to rectify this behaviour it will be important to find out why the cat is reluctant to use his litter facilities or to go outside to toilet in the garden.

A bit of privacy, please

This cat has selected a quiet and secluded location so that he can go to the toilet in peace. The soft rakeable soil allows him to dig a small indentation in which to deposit the faeces. Then he will rake over the hole to ensure that the excrement is safely buried to reduce the risk of infection or parasite transfer.

Health check

Urine deposits in the house may be the result of toileting inappropriately or territory marking. When the deposit is on a horizontal surface and in a quiet location, it is more likely to be an inappropriate toileting behaviour. This cat appears to be slightly hunched, which could be indicative of pain, therefore it will be important to make sure that this cat is in good health and that there is no underlying medical cause for the behaviour.

Addressing problems

Toileting in inappropriate locations in the house can lead to a rapid breakdown in the relationship between cats and their owners. The behaviour is perceived as both odd and unacceptable and sadly it can lead to decisions to rehome or even euthanase cats. This is particularly sad because the causes of inappropriate toileting are often very simple to rectify with a basic understanding of the feline requirements for a suitable latrine.

I don't want to eat here

The location of a litter tray can affect your cat's response to it. He looks for a quiet and undisturbed spot and will avoid toileting close to where he eats. The positioning of food bowls close to a dirty tray such as this one is not only unhygienic but also very distressing for a clean creature such as a cat. Feeding stations and latrines need to be placed completely apart and both need to be positioned in quiet places so that your cat can both eat and toilet in peace.

TEXTURE ISSUES

Cats need a soft, rakeable substrate in which to toilet. Their instinct is to bury urine and faeces in order to decrease the risk of spreading infection and parasites and of attracting predators. Hard pelleted litters can be associated with problems of toileting outside the tray because they can be difficult to dig in and may have an added disadvantage of being uncomfortable for the cat to stand on!

Keep it clean

Cats are fastidious creatures and daily scooping out of deposits from the litter tray is essential to ensure that it will continue to be used. Once a week the tray should be emptied, cleaned with boiling water and fresh litter provided. If the tray is full of old deposits cats will look for an alternative location and this is a common cause of indoor toileting problems.

Not big enough

This small kitten is only just able to fit into this tray and before long it will not be big enough to allow him to comfortably stand up, turn round, rake and squat. This small tray also has far too little depth of litter and it isn't possible to dig an indentation that is deep enough to cover up faeces and urine deposits satisfactorily.

Multiple cats, multiple toilets

Cats need to toilet in peace and when there are two trays positioned side by side you are only creating one latrine site. If the cats are not compatible this can be a source of stress, since each cat needs free and immediate access to the trays when required. Moving the trays into two separate locations will solve this problem.

MARKING

Scratching

Cats are sociable creatures but their ultimate survival is down to them as individuals and they are keen to avoid confrontation and injury. They will try to evade any non-essential encounters with other cats and have an elaborate communication system to help them to achieve this. Part of that system involves the use of various marking behaviours. These enable our cats to leave messages that can be read by other cats without the necessity for face-to-face encounters.

Ooh, that feels good

Our cats seem to derive great satisfaction from a good scratching of their claws. This will reveal glistening new claws beneath, ready for the kill. While he is scratching the cat is also depositing scent from glands on his paw pads and marking out his territory.

Just what I need

Cats living in a domestic setting still need ample opportunity to scratch. Some will get this chance when they are out in the garden but for others purpose-made cat scratching posts can be an important outlet for this behaviour within the house. This cat is making good use of his post and the positioning of the scratched surface over halfway up the post highlights the need for it to be tall enough to enable effective scratching to occur.

Hang on

This cat is scratching at a curtain but this is more likely to be a case of climbing than primary scratching. The curtain is not sturdy enough to enable the cat to strop his claws successfully but the material is easy for sharp claws to cling onto and as the cat climbs he will inevitably scratch the fabric! Providing a cat aerobic centre would help with this problem as the cat would have the posts to scratch on and the platforms to climb onto.

Now I can reach

When cats scratch they need to get a good purchase on the surface and this involves stretching out their bodies to their full extent. This cat is using the bird bath to give him the correct base from which to stretch himself up this tree trunk, which offers an ideal surface to dig his claws into.

Rubbing

When cats rub their faces against objects and people the behaviour is often seen as endearing but baffling. Humans are largely unaware of the scents in the world around them but cats communicate in the olfactory dimension. They produce signals from glands all over their body and the ones on the face and flanks provide reassurance for cats and help to consolidate their relationship with their environment.

This is mine...

Marking by rubbing will take place inside the home and this cat is checking out the scent of this new chair and marking it with his own scent from glands on his face and body. This will help to incorporate the chair into the scent profile of the home and therefore establish that it belongs there.

You're mine too!

Rubbing behaviour is not limited to inanimate objects and as this owner returns home the family cat greets her with a rubbing action on her legs, which enables the cat to mix his scent with that of his owner. This rubbing action is designed to establish the shared communal scent that identifies the owner as part of the cat's social group.

This is where I live

By rubbing his face on the fence post this cat is depositing important scent information from the glands that produce the facial pheromones. These signals help to identify the post as part of this cat's territory and make him feel safe and secure. They also signal to other cats that the area is occupied.

Middening

Using faeces to communicate is bizarre from a human point of view! We like the fact that cats bury their toilet deposits and keep the place tidy and when faeces are found lying out in the open owners are often embarrassed. However, these deposits are intentionally left uncovered as they are being used to convey information to other cats about territory boundaries. The problems happen when humans rather than cats find the faeces and disputes with our human neighbours can sometimes arise as a result.

I was here

Middening is a marking behaviour using faeces and deposits that can sometimes be found in very unusual places such as the top of sheds or outbuildings. This cat leaves small deposits of faeces on the ridge of this roof. The scent of the faeces wanes over time so the deposits can tell other cats using this territory how long ago this cat was there. This information allows the cats to time-share the territory and avoid unnecessary encounters.

Cats do not respect boundaries between gardens and the presence of middens on lawns and shed roofs can rapidly lead to neighbourhood disputes. Making your garden more attractive from a feline perspective and offering your cat increased signs of security at home can help to reduce their need to mark out their territory in this way.

Keep out

Cat flaps can be a significant threat to the security of a timid cat. This cat has deposited a pile of faeces on the brown rug, just out of shot. From a feline perspective the deposit sends out an unmistakable message of occupancy to any potential intruder.

This is my patch

When cats feel vulnerable in their own gardens or their outdoor territories they may use a midden to decrease the risk of confrontation. Leaving small deposits of faeces in the middle of this expanse of lawn is one way for him to signal to his feline neighbours that he is still in residence.

Urine marking

One of the most difficult behaviours for owners to tolerate is urine marking because the smell of urine is highly offensive to people! However, cats deposit urine as a means of communicating with other felines and increasing their own feelings of safety and security. Depositing urine inside the house often has the opposite effect of making the environment more hostile as frustrated owners use punitive techniques to reduce this unwanted behaviour.

Interpretation

As cats walk through their territory they check for scent messages from other cats. This method of communication ensures that the sender and receiver of the message do not need to be in the same place at the same time and avoids the possibility of unwanted confrontation. This cat is sniffing intently at a message but shows no sign of fear or apprehension. The message is purely factual and not threatening. The intense sniffing behaviour is part of the flehmen response, which enables this cat to pull the scent information from the mark into a specialized organ above the hard palate, which is needed to fully interpret the information conveyed by the urine deposit.

Just doing my rounds

Urine spraying can also be used for cats to reassure themselves. This cat sniffs at the cupboard door with interest. Previously it has deposited urine on the door and it is checking to make sure the message is still fresh. The aim of the urine is to make the cat feel more secure and his body language is calm and controlled.

Leaving a message

When urine is used as a marker the cat adopts this characteristic body posture and reverses up to a spraying surface, such as this bush. The tail is held upright and the cat stares in a concentrated fashion as it sprays a jet of urine onto the vegetation behind. The deposit is at a convenient height for other cats to read it as they pass by.

THREATS TO SECURITY

Domestic unease

Cats are territorial creatures and the security of the place where they live is very important to them. Their territory can be divided into three zones. The outer hunting territory and the middle home range will be shared by a number of cats. In contrast, the inner core territory is only shared with other cats from the same social group. It is very important for cats to feel safe and secure in their core territory.

Let me eat alone

Eating is an important activity in the core territory. Cats are solitary feeders and their natural instinct is to eat alone. If food is provided in locations that encourage more than one cat to be interested in the same food source, such as the one shown here, it can lead to considerable levels of stress and tension.

This is my space

The compatibility of cats living in the same house is an important issue. If cats that are hostile to one another are forced to share the same home it will lead to long-term stress. This can make the cats feel tense and uneasy in their core territory, which should be their place of security and solace.

I'm watching you

If there is any tension within the core territory it is important that cats are able to retreat to a safe hideout while they observe the situation and judge whether it is safe to come back out. This cat is using a purpose-built hole in the skirting board to observe from a position of safety. If the bolt-holes are being used regularly it will be important to find out why the cat does not feel safe and work to rectify that situation.

Friends

Cats can make ideal pets for small children but it is important to make sure that any interaction between the two is supervised and is beneficial to both. In this photo the child is approaching slowly and the cat is not showing fear but if children are too loud or boisterous in their approach it can make the cat frightened and decrease its sense of security.

Life in a goldfish bowl

Owners often wonder why a cat will stare out of the window for hours on end, sometimes thrashing their tail back and forth as they do so. Observation of his territory is an important part of your cat's natural behaviour repertoire and in situations where there is a perception that the territory may be under threat, the degree of visual access into and out of the house can be very important.

Creating privacy

If it is necessary to obscure a cat's visual access to the outside world, curtains are not usually successful. Cats can pull them down or just sit the other side of them. Instead, temporary frosting material can be applied to the window while the owner works to increase the cat's sense of security at home and maybe does some garden design work to obscure the visibility into the house for passing cats.

Stop looking

Fence posts, shed roofs and even cars parked in driveways can all offer perching places for other cats which can enable them to stare into houses and visually threaten the cats living inside. If the resident cat is showing behavioural problems as a result of this visual intimidation, then strategic positioning of plant pots and garden furniture can be used to obscure the view and make the core territory safer.

Let me at him

Even when a core territory is physically secure, cats can feel vulnerable if there is a lot of visual access into their home from the outside world. If the cat can see other cats traversing his territory but cannot gain access to the other cat or communicate with it effectively there can also be an increase in frustration.

Cat flap conundrums

Most owners think that free access to the outside world is what their cats crave and cannot understand why their pet will not readily use the cat flap they have bought and installed. Teaching some cats to use the device can take patience and we need to understand that a hole in the home's defences is not always seen as a positive thing from a feline perspective.

Is it safe?

Most modern-day cat flaps are made from transparent material and this can present some problems by increasing visual vulnerability. This cat is staring out through the flap and appears to be hesitating before going outside. This could be due to the presence of another cat on the other side of the flap. Other cats in the neighbourhood may have previously sat staring in through the flap and this makes the resident hesitant to use this point of entry and exit.

I'm back!

Cat flaps are seen by most owners as a positive invention and when cats use them successfully they certainly can be. They enable the cat to control its own access to the outside world and give a degree of freedom. This young kitten has already got the hang of using the flap and is confidently returning to his home.

Come on through

Some kittens struggle to learn how to work the flap and some may be fearful of it in the early introductory stages. It can help for the owner to stand on the opposite side of the flap with a bowl of tasty food to encourage the kitten to push the flap to get through to the food.

Too close for comfort

Cat flaps provide both a visual and physical point of entry into your cat's core territory and can therefore make his home environment less secure. It is important to consider this when positioning the flap in relation to litter trays or feeding stations. This litter tray has been placed close to a transparent cat flap. Although this young kitten seems happy to use the litter tray it is unlikely that an adult cat, with a more developed perception of potential threat, would feel safe enough to toilet in this particular location.

SOCIABILITY

Multi-cat issues

Cats are social creatures and can live happily in groups of compatible individuals who are often related to one another. They do not have a pack social structure like dogs and their survival is the responsibility of each individual. As a result, relationships between cats can be fragile and they have very limited behavioural strategies for dealing with conflict. Living alongside other cats, either in the same home or in the neighbourhood, can pose some significant challenges.

Cat fight

Overt aggression within feline households is not common but when it happens it can be very severe. At this stage it can be difficult to resolve the situation and these cats may quickly progress to the point where they will fight on sight. Lack of ability to use natural feline behaviours, such as hiding and getting up onto high resting platforms, in order to maintain distance from one another will make the incidence of fighting more likely.

We're friends

When cats are related to each other or are reared together from an early age they can develop close friendships and enjoy each other's company. These two cats live in the same household and can easily relax in close proximity to each other.

Tension

Even when there is no actual fighting in a multi-cat household there can be considerable tension between incompatible cats that are forced to live under the same roof. These cats are using a combination of body posturing, visual communication and slow movements to indicate their desire to avoid outright confrontation. However, the chronic stress from the social tension between them can have serious behavioural and medical consequences for both cats.

Day-to-day interaction

All owners want their cats to get along with one another and when they do not display overt aggression to each other it can be tempting to assume that all is well. However, it can be difficult to gauge the nature of relationships between cats because they tend to avoid confrontation whenever possible. It is therefore important to watch for subtle signals of tension between feline housemates and avoid forcing cats into situations where they have to tolerate unwanted social interaction.

Can I eat in peace

It is common practice in multi-cat households to feed cats side by side but it is highly stressful for the cats. The cat on the right stops eating and observes his housemate. The cat on the left continues to eat but shows a tense body posture combined with wary eyes and alert ear positions that are consistent with a state of anxiety.

That's annoying

An adult is more likely to tolerate the arrival of a kitten than an adult feline stranger but sometimes the play of a young kitten is not compatible with an older cat. While this kitten attempts to use the cat's tail as a plaything the older cat shows signs of tension in his facial expression. The rotated ears, the forward-facing whiskers and the staring eyes all indicate stress!

Too many cats...

Food is a survival resource and if it is only available in one location at a set time in the day then all of the cats in the household will have to be there to receive it. However, feeding is a solitary affair in feline circles and although these cats will override social tension in order to be in the kitchen ready for feeding time, their body language, pacing behaviour and vocalization signals indicate that they find this feeding ritual stressful and difficult to cope with.

I'm safe up here

When social tension exists within a multi-cat household every individual needs the ability to get away from the stress. This cat has retreated to the top of furniture in the living room and from this vantage point he observes the activities of the other cats. It is important to allow this cat to come down in his own time and under no circumstances should the owner attempt to bring him down from his safe bolt-hole before he is ready.

The neighbours

It can be fascinating to watch your cat communicating with a neighbourhood rival but often the full meaning of their interactions can be difficult to comprehend. Cats use a subtle combination of body language, vocal signalling and smell to get their message across and learning about these signals helps humans to understand more about the tension that can exist in multi-cat neighbourhoods.

I didn't see you there

When cats encounter each other unexpectedly there is more likelihood of aggressive signalling. This Siamese is hissing in an attempt to avoid contact but its ear position and crouched rear end body posture suggest that it is far from confident. The Persian is startled by this behaviour and prepares for a potential confrontation. However, both show a weight distribution that suggests that a slow and stealthy retreat is their favoured course of action.

EAR PROTECTION

Cat fights invariably result in injury as claws are protracted and bites are not inhibited during these interactions. Hearing is necessary for the location of prey and therefore the protection of ears during confrontation is important. When the cat is confident in its encounter the ears may be erect and pulled back against the head but in situations of fear the ears are more likely to be flattened to the head.

Passing through

These feline neighbours are travelling through a shared piece of territory and in order to avoid a confrontation the ginger cat uses a combination of body posture, slow movement and possibly vocalization to signal to the tabby that he is just passing through and has no intention of stopping!

Fight, fight!

When passive communication fails, neighbouring cats can become involved in physical fights. These two cats are now highly aroused and the aggression can quickly escalate. Paw swiping is usually the preferred initial approach as it enables the cats to maintain some distance from each other, but once they have come into physical combat, as seen here, the likelihood of resolving this dispute without physical injury to both parties is small.

Watching and waiting

Sudden bursts of feline activity sometimes take owners by surprise – one minute your pet seems to be quietly watching the world go by and the next he is grabbing at your shoe laces! Cats are programmed to notice change in their environment, and sound and movement are particularly interesting to them. They will watch patiently for some slight activity, then take their opportunity to pounce.

Don't rescue me!

It is perfectly normal for cats to use elevated vantage points to observe their territories and climbing up trees or onto the tops of kitchen cupboards are normal feline behaviours. However, sights such as this of cats high up in the branches of a tree will often induce worry and concern in well-meaning people who then attempt to use ladders and even the fire brigade to bring the cat down. Such rescue activity is unnecessary but it also runs the risk of being misinterpreted by the poor cat as threatening intervention.

You can't escape

It is very important to channel natural feline behaviours, such as predation, into suitable and acceptable outlets and providing appropriate toys for kittens from an early age is an important part of this process. This kitten shows focus and determination as he attempts to capture and dispatch this toy mouse and this will help to reduce unwanted predatory advances towards people.

They're alive!

Any rapid and unexpected movement can induce an instinctive predatory response in cats from an early age. This can result in some amusing behaviours. These kittens are reacting to the presence of shoelaces that move erratically when the person walks. They grab at the laces with their paws and claws and watch the laces intently for their next move.

Aggression to humans

When cats show aggression to humans we are often perplexed by their behaviour and may even label them as 'evil'. The truth is that cats are programmed to avoid conflict and when they show aggression it is a sure sign that something is wrong. Understanding more about their view of the world will help us to make sense of such responses and recognize that feline aggression is usually triggered unintentionally by something that we humans have done.

Don't come closer

This cat is fearful about the approach of a person and is showing defensive ear posture and hissing vocalization which clearly indicates his perception that he is being threatened. The front paw is outstretched and could be getting ready for a swipe with extended claws. However, at this stage he is giving a convincing display which will hopefully result in the person retreating without any need to escalate the situation.

Ouch!

Human handling can sometimes be misinterpreted by cats and when we stroke them too rapidly or vigorously their instinctive response can be to close in on the arm and hand with all four feet to prevent it from moving again. This may be followed with a bite of the wrist as seen here. The person has responded appropriately by letting her hand and arm go limp. As the cat realizes that the 'threat' has subsided this potentially aggressive encounter can be defused.

I'm brave

Cats are well armed with claws which enable them to swipe out toward perceived threats and, because the cat is still able to maintain a safe distance, the behaviour may look confident, even though it is driven by fear. Watching other body language signals such as vocalization and ear position help us to identify the underlying negative emotion as can be seen in this photo.

AT REST AND PLAY

Cats sleep anywhere

Cats naturally engage in short bursts of energy-consuming activity and then spend long periods of time sleeping. Rest and play are therefore important aspects of a cat's daily activity and these behaviours must be catered for when we live with cats as pets. It is necessary to ensure that resting places are easily accessible and readily available and that play sessions are frequent and fulfilling.

Still checking

Warmth and comfort are important features of a suitable resting place and this cat has chosen the soft material of the throw on the back of the sofa. He is snuggling down into his chosen safe haven but is not yet relaxed enough to stop scanning his environment with his ears.

This is comfy!

Republished in the 1990s, Eleanor Farjeon's poem 'Cats sleep anywhere', celebrates the cat's ability to find appealing locations to curl up and take a nap, and it is certainly true that cats often use multiple resting areas within the home. Even when owners have gone to the trouble of buying expensive cat beds their pets will often choose to curl up in the washing basket instead.

I love books

In order to relax completely cats will often seek resting places elevated off the ground and that offer privacy and seclusion. This cat is almost hidden on the bookshelf among the antique books and his upturned paw and loosely dangling front leg are clear signals that he feels completely safe and protected.

It's playtime

It can be frustrating when your cat seems to be more interested in an elastic band or a screwed-up piece of paper than the lovely toy that you bought for him from the pet shop. We need to remember that cats view the world differently from us and that what we find attractive may not hold the same appeal for your pet. Understanding the qualities that make toys interesting from a feline perspective will help you to get the most out of playtimes.

Hours of fun

Although play is more readily associated with kittens it is an important part of the daily routine for adult cats as well. It is therefore important to maintain a supply of suitable toys throughout the kitten's development and into adult life. Toys do not need to be expensive. This suspended cork has many of the important features of a successful cat toy, including unpredictable movement and an interesting texture to grip with claws. Its value could be enhanced by colouring it with fluorescent camouflage.

I can make it move

Fluorescent colours can help to make a toy look different from different angles. This brightly coloured ping-pong ball combines visual appeal with the hollow sound it makes as it bounces, thus keeping this Siamese kitten fascinated. By patting the ball with his paw the kitten can make it move, maintaining his interest for even longer.

This is interesting

The cat's sensory system is designed to detect minute changes and play items that have unexpected movements and high-pitched sounds are likely to be popular. This ball has a bell inside and the sound is attracting the interest of this kitten. If the kitten can be brave enough to pat at the ball with its paw, the holes in the surface of the ball will give it an erratic movement that will make it even more interesting.

This feels good

Play items that are small and light enough for your cat to pick them up and carry them around can add a further predatory dimension to the play experience. This kitten is holding a small drum of catnip powder in his front paws and biting at it and dribbling in response to the catnip. Not all cats would react in this way, as only 50 per cent of cats carry the gene that enables them to respond to and enjoy the drug-like properties of the catnip plant.

Pay attention

Playing with your cat is so much fun but as he gets older it can sometimes appear that he is losing interest. Just as you are settling in for a pleasurable playtime with your pet you find that he gets up and walks away! These short bursts of activity are perfectly normal for cats and it is important to play with your cat, however short the session.

Keeping my interest

Provided that a toy offers enough variation to maintain interest then cats can play quite contentedly on their own. This kitten is happy to amuse himself using a toy that combines different textures with the unpredictable movement of the coloured feathers and the rattling sound caused by the grains of rice in the body of the mouse!

Get me down!

Cat activity consists of short bursts of high energy-consuming activity, which is sometimes misperceived as a bout of feline madness – that crazy five minutes. During these episodes young cats often find themselves getting into difficult, but amusing, situations. This kitten has climbed up these flimsy curtains but his claws have become entangled in the thread, so this poor individual is likely to be hanging from the curtain for a little longer than he intended!

I've had enough

Feline play is fairly short-lived and a session of one to two minutes should be considered a success. If this little girl is no longer able to maintain her cat's interest in the wand-style toy, it is time to end the play session and try again later. Playing little and often is far more beneficial for the cat than persevering once his interest is lost.

It's never ending!

Young kittens are full of mischief and while they do not sustain play over long periods of time they can certainly create a lot of havoc in a short play session! The toilet roll is particularly rewarding for this kitten because each time he grabs at it, it will unravel a little bit more.

Curious cats

Cats are renowned for their curious nature! They spend
a considerable proportion of their day exploring the
environment but sometimes this can drive them to
behave in ways that we find both amusing and bizarre.
Anything that looks different, moves rapidly or sounds
intriguing will be worthy of investigation. Inquisitive cats
are particularly drawn to going through apertures and
openings to see what lies on the other side!

Oops, I'm stuck

When this kitten set out on one of his
daily voyages of exploration he came
across the interesting texture of this
curtain. His interest in the material led
to exploring it with his paws but his
intention of running up the curtain has
been thwarted, as his front claws have
become embedded in the curtain.

A quick spin

Sometimes the old adage 'curiosity killed the cat' seems all too true. This small kitten has jumped in through the open door of the tumble dryer and the result is a very cute photograph of a much-loved pet. However, this sort of behaviour could end in tragedy and there have been reports of cats climbing inside household appliances without their owners being aware. Make sure you check inside your machines before using them, and keep the doors of your appliances closed when you are not at home.

What's in here?

Cats are naturally inquisitive and when this individual sees a paper bag on the floor he cannot resist the temptation to peep inside. As long as the cat is not nervous or startled by the crinkling sound of the bag it is likely that the sound and movement will sustain his interest and encourage the exploration game to continue.

HOME AND GARDEN

Home sweet home

Territory is important to cats. While the hunting and home ranges may be shared with other cats, the core territory should offer the opportunity for periods of solitude and seclusion. Cats will naturally tolerate the presence of only their own social group within this important area. The three main activities within the core territory are eating, sleeping and playing.

A quiet dinner at home

Eating is an important activity within your cat's core territory and helps to identify this location as his secure home base. It is important that your cat is able to eat alone and truly relax, so it is best to place the food bowl in a quiet area where he will not be disturbed.

Sunbathing

In the safety of his core territory this cat can totally relax in the sunshine, in the knowledge that he will come to no harm. His level of relaxation can be gauged by his body posture and the fact that all of his four feet are no longer in contact with the windowsill. This makes retreat difficult to achieve at speed and thus signals a high level of confidence and relaxation.

Playing at home

Along with eating and sleeping, it is essential that your cat also experiences secure play activities within his core territory. The provision of stimulating toys that retain his interest will ensure that your cat plays regularly within his safe home zone. This activity tower offers a central scratching post and a number of other features that are designed to induce and maintain feline play, such as the suspended, brightly coloured ball and the protruding rodent heads.

Sharing territory

If you live in an urban area your cat will probably find himself sharing his home range with several other neighbouring felines. Observing, marking and defending his territory will therefore be important activities and can take up a large proportion of his day. Sometimes, when your cat is on surveillance duty, he will appear to be just whiling away the hours; actually he is doing a very important job!

My kingdom

The home range of any cat is shared with his feline neighbours. It can therefore be helpful for your cat to find elevated resting perches in the garden that enable him to observe this shared territory, without running the gauntlet of direct physical encounters. Perching sites should only be big enough for one cat, like the one shown in this photograph. However, you also need to consider the welfare of any birds nesting in the box, if your cat takes this up as his preferred position in the garden!

Eye spy

When kittens are small, they find safety in numbers. In this picture the two kittens use an elevated vantage point to monitor activity in their outdoor territory. Their body language – alert ears, eyes and whiskers – suggests that there is plenty for them to watch.

FINDING A LATRINE

It can become challenging for cats in multi-cat neighbourhoods to find secluded locations in which to toilet. Even when a suitable location is found, over-population in urban areas can mean that gaining access without having unwanted interaction with other cats can be a factor in some cases of indoor toileting. It can be beneficial in these cases to provide latrine sites close to the house that can be accessed more readily.

Keep out of my way

When cats need to walk across shared territory there is the risk of unwanted encounters with other felines. These two cats live in the same household but are not socially compatible. When they are both in the garden at the same time hostility can be obvious but they still try to avoid out-and-out physical fights. In this photo you can see the use of effective vocal communication combined with the necessary weight distribution to ensure rapid get away should it become necessary.

Unwanted hunting

Unwanted hunting behaviour can seriously damage the cat–owner relationship. Even when we appreciate that our pet cat's behaviour is entirely natural, it can be hard to accept the sight of dead rodents or birds deposited on our kitchen floor! If your cat is particularly successful in his hunting behaviour you may need to control your cat's access to potential prey, especially at times – such as their breeding seasons – when birds and small mammals are particularly vulnerable.

Fly away quickly!

Birds often have a special place in our hearts and many people find the thought of cats killing small birds distasteful. However, hunting is natural behaviour and it is important not to actively punish him, no matter how annoying it is. It will be far more effective to spend your energy on making it difficult for your cat to get access to birds, especially when they feed and breed. However, some individuals, such as this kitten, can be ingenious and determined to find ways of reaching their potential prey.

The hunting range of the territory is shared by a number of cats and, in order to avoid meetings and confrontation, they use a system of time sharing. Ensuring that the territory is used efficiently is in the interest of all the cats and the marks are purely informative and not threatening. When cats read the signals left by their fellow felines they do not show any signs of fear but simply alter their route through the territory to avoid unnecessary encounters.

Nice and tasty

Hunting is a feline behaviour that induces very mixed responses from owners. On the one hand the sight of this cat consuming a young rodent may be repulsive to some people but the thought of the cat controlling the rodent population in the neighbourhood is probably going to be seen as a positive thing.

Gone fishing

When this cat is fascinated by the contents of this garden pond the most pressing question is 'What is the pond stocked with?'. If there is the risk of him fishing out some ornamental fish then the problem suddenly moves up a notch in terms of severity, and owners can find themselves faced with large bills for the cost of replacing the contents of the pond.

THE ODD AND OLD

Strange behaviour

While much feline behaviour can be explained via an understanding of their natural instincts and their emotional responses, there are some feline activities that seem to defy understanding and can only be defined as odd or bizarre. Old age can also influence behaviour and some special considerations may be necessary when caring for an elderly feline companion.

Chasing nothing

When your cat is spending time chasing images that are not actually there it can seem amusing but hunting imaginary mice and catching imaginary flies are behaviours that can be associated with some medical conditions. It is important not to interrupt the behaviours, especially with any form of punishment, as this can lead to problems of frustration. The best approach is to talk to a veterinary surgeon and ask for the behaviour to be investigated in more detail.

I can't catch you

This cat is fascinated by the light source on the wall and is chasing it intently. This sort of behaviour can be amusing for us to watch and it is tempting to use the reflection off the surface of a watch or the light from a laser pen to encourage your cat to follow a light trail. The only problem with this form of play is that it can increase your cat's interest in the imaginary object and may induce behavioural problems, such as compulsive light staring or frustration because the light source is never caught.

Can I eat this?

Wool eating is probably one of the most bizarre feline behaviours but is perhaps more accurately referred to as fabric eating or 'pica', which means an abnormal appetite leading to the ingestion of non-nutritional substances. This behaviour is more commonly seen in oriental breeds and these two Siamese kittens are showing abnormal interest in the jumper. The wool eating may start as a sucking response rather then chewing and ingesting and the development of the behaviour can be complex.

Just being fussy

The way in which cats eat and drink can sometimes be perplexing. They may turn their nose up at the food or water that we provide and we assume that they are just being picky. However, these responses could be a sign of emotional pressure and it is important to watch your cat's behaviour closely to determine the reason behind it. Trying to 'think' cat can help us to make sense of these foibles.

Drop by drop

Many cats are attracted to moving water sources and will shun their water bowl in favour of a dripping tap. A more sophisticated way of providing running water for cats is to use one of the commercially available water fountains. The most important thing for all cat owners is to ensure that their cats drink enough water; if a dripping tap encourages this, it is certainly worthwhile.

Refusing food

When a cat refuses to eat food that is offered it is important to rule out medical causes but also to consider potential emotional reasons for this behaviour. Chronic stress in multi-cat households can lead to a reluctance to eat and anything that makes the cat feel vulnerable in the area around the food bowl could also be significant.

Tasty grass

If cats do not have any access to the outdoor environment it can be helpful to provide them with some green vegetation, in the form of indoor grass plants. It can also be beneficial for those individuals who are responsive to catnip to grow some of these plants and give the cat the opportunity to take advantage of the drug-like properties of the plants when the need arises!

Elderly cats

There is something very special about the relationship with an elderly cat but some of what it does can seem a little quirky. The ageing process can lead to problems of confusion and disorientation and bizarre behaviours can emerge as a result. The onset of age-related medical conditions such as arthritis can also affect its actions. Pain in cats is primarily associated with changes in activity level and temperament rather than lameness, so it is important to be on the lookout for these changes as your cat ages.

I can't see

As cats grow older, owners may notice a significant change in their pets' behaviour. The first thing to do in these situations is to take the cat to a vet to check for medical problems relating to the senses, since blindness and deafness are both common in the ageing cat population. If cats are blind, such as this elderly individual, it may be necessary to avoid sudden changes in the layout of the house, but generally cats adjust remarkably well to their loss of sensory information.

It hurts to walk

This ageing cat and its owner share a special bond and the cat shows no signs of apprehension related to being picked up and cuddled in this way. Under normal circumstances cats are reluctant to be lifted off the ground and would prefer to walk under their own steam so, in cases where older cats seem unusually contented to be carried, it would be sensible to check that there is no pain from arthritis that the cat could be avoiding by allowing himself to be carried.

I'm confused

This cat is vocalizing in a random manner and the sound is associated with pacing behaviour that is repetitive and purposeless. Disorientation is one of the main signs of dementia in cats (feline cognitive dysfunction). Persistent but non-focused vocalization is also a symptom of this condition and it can cause problems for owners, especially when it is associated with night-time waking. It is important to take these cats to the vet, as there are medications available now that can significantly help with managing the condition.

INDEX

ACKNOWLEDGEMENTS

It is impossible to write a book without the help and support of many people. Firstly, I would like to thank Trevor Davies and Ruth Wiseall at Hamlyn for their patience and understanding. Grateful thanks are also due to all of the staff at my veterinary practice for their support and for keeping the practice going when I have been busy finishing this book! I would like to thank my Flatcoated Retriever, Cira, who sadly died unexpectedly during the writing of this book, for giving me unconditional love and understanding. Finally, a huge thank you goes to all the cats I have known and owned and in particular to Muffin and Timbit for giving me the privilege of sharing their lives.

Executive editor **Trevor Davies**
Editor **Ruth Wiseall**
Executive art editor **Penny Stock**
Designer **Ginny Zeal**
Picture researcher **Ciaran O'Reilly**, **Zoë Spilburg**
Production controller **Marián Sumega**

PHOTOGRAPHIC ACKNOWLEDGEMENTS
Key: **a** above, **b** below, **c** centre, **l** left, **r** right

Alamy Andrew Linscott 70; Antje Schulte 93 b; blickwinkel 19 b; Bruce Coleman Inc 32; David Askham 13 a; David Rowland 84; dk 81 b; Frisk 88; Geoff du Feu 87 a; Gerry Pearce 74, 75 a; Helene Rogers 11 a, 46; imagebroker 10; Isobel Flynn 63 b; Juniors Bildarchiv 12, 20 a, 47 b, 56, 61 a, 67 ar, 73 b, 78; Jupiterimages/Brand X 83 a; Kerrie Snowdon 61 b; L Kennedy 77 b; Peter Cavanagh 60; Petra Wegner 16; Rob Walls 53 b; Robert Estall Photo Agency 69 b; Rodger Tamblyn 57 b; Sigitas Baltramaitis 73 a; Steppenwolf 92; superclic 15 a; Wildlife GmbH 55 b. **Ardea** Ardea London 25 b, 43 a, 45 b, 69 a; Brian Bevan 87 b; Francois Gohier 44; Jean Michel Labat 9 b, 23 c, 30, 31 b, 41 b, 43 b, 72, 83 b; John Daniels 9 a, 17 a, 18 b, 19 a, 29 a, 47 a, 49 c, 53 a, 77 a, 80, 82; Rolf Kopfle 15 b, 50, 85 a. **Art Directors and Trip** Helene Rogers 67 al. **Corbis UK Ltd** DK Limited 33 a, 35 b, 41 a, 42, 71 b, 89 b; Simon Marcus 79 c; Son Desert 31 a; Tracy Kahn 79 b. **DK Images** Jane Burton 65 b. **FLPA** imagebroker 8. **Fotolia** Simone van den Berg 40. **Getty Images** Bread and Butter 93 a; Chris Rose 55 a; Dea D Robotti 28; Felipe Rodriguez Fernandez 7; Glow Images 39 b; Jane Burton 21 b, 52; Jean Paul Nacivet 38 b; Jesse Burke 54; Sami Sarkis 77 c; Toby Maudsley 89 a; Todd Gipstein 14. **istockphoto.com** Joanne Harris and Daniel Bubnich 26; Martin Carlsson 49 a; Sara Robinson 17 b. **Masterfile** Nora Good 75 b. **Nature PL** Jane Burton 63 al. **Octopus Publishing Group Ltd** Steve Gorton 63 ar; John Daniels 51 bl. **Photolibrary Group** Beanstock Images 65 a; Carrie Villines 39 a; Index Stock Imagery 4, 22; Juniors Bildarchiv 2–3, 11 b, 18 a, 21 a, 23 b, 25 a, 34, 37 a, 45 a, 64 , 66, 68, 76, 79 a, 81 a, 91 a; Konrad Wothe 71 a; Layer Layer 38 a; Martin Rugner 27 a; Motor Presse Syndication 90; Oxford Scientific 24; Paolo Curto 59 b; Radius Images 91 b; Santa Clara Photononstop 58; Stephanie Deissner 23 a, 35 a; Todd Bannor 59 c; Werner Pokutta 20 b; Yoshio Tomii 86. **Photoshot** Lacz Gerard 59 a; NHPA 85 b; Servanne Sohier 33 b; Vicente Roca Gisbert 48, 62 b; Westend61 5, 13 b, 27 b. **RSPCA Photolibrary** 51 a & r, 29 b. **Warren Photographic** 1, 6, 29 c, 36, 37 bl & br, 49 b, 57 a, 66–7.

My Favourite Dogs

GOLDEN RETRIEVER

Jinny Johnson

W

FRANKLIN WATTS
LONDON • SYDNEY

An Appleseed Editions book

First published in 2013 by Franklin Watts
338 Euston Road, London NW1 3BH

© 2012 Appleseed Editions

Created by Appleseed Editions Ltd,
Well House, Friars Hill, Guestling,
East Sussex TN35 4ET

Designed and illustrated by Hel James
Edited by Mary-Jane Wilkins

ISBN 978 1 4451 2181 9

Dewey Classification: 636.7'527

A CIP catalogue for this book is available from the British Library.

Photo acknowledgements
t = top, b = bottom
page 1 iStockphoto/Thinkstock; 3 Jagodka/Shutterstock; 5 Joop Snijder jr./
Shutterstock; 6 iStockphoto/Thinkstock; 7 Hemera/Thinkstock; 8-9 Lisa A.
Svara/Shutterstock; 10-11 iStockphoto/Thinkstock; 12 Comstock/Thinkstock;
13t iStockphoto/Thinkstock, b Mat Hayward/Shutterstock; 14 amidala76/
Shutterstock; 15 Bernd Obermann/Corbis; 16 Apple Tree House/Thinkstock;
17 Tom Nebbia/Corbis; 18-19 Ryan McVay/Thinkstock; 20 iStockphoto/
Thinkstock; 21 Michelle D. Milliman/Shutterstock; 23 tstockphoto/Shutterstock
Cover Onur ERSIN/Shutterstock

Printed in China

Franklin Watts is a division of Hachette Children's Books,
an Hachette UK company.
www.hachette.co.uk

Contents

I'm a golden retriever!

I'm a friendly, happy dog
and I'm loyal to my owners.

I love to please people
so I'm very obedient,
but I like to play as well.

What I need

I'm quite big and I like plenty of exercise. I love to swim, too. I like being part of the family and I'm not a good guard dog because I'm so friendly!

Best of all, I love to bring back sticks and balls if you throw them for me.

The golden retriever

Double-layer coat; outer layer straight or wavy

Strong tail

Colour:
cream to golden

Height:
51–61 cm

Weight:
25–34 kg

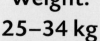

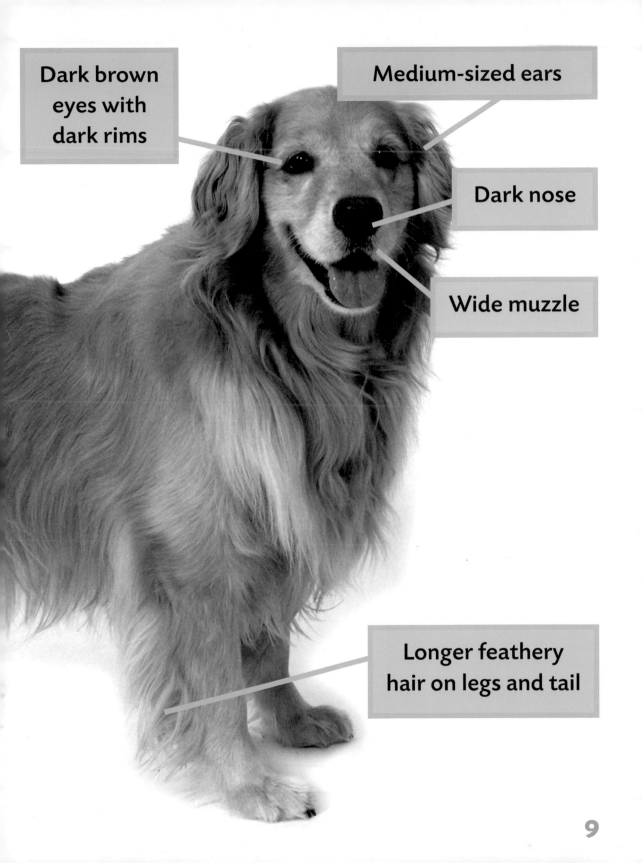

Dark brown eyes with dark rims

Medium-sized ears

Dark nose

Wide muzzle

Longer feathery hair on legs and tail

9

All about golden retrievers

This dog was first bred in Scotland to fetch (retrieve) birds, such as ducks, shot by hunters.

Don't worry, this is a pretend duck. The dog is being trained to bring things back to its owner.

A golden retriever can hold objects very carefully in its mouth and bring them back to its owner.

Growing up

Golden retriever puppies need to be with their mum until they are about eight weeks old. When you take your puppy home, she may be frightened at first.

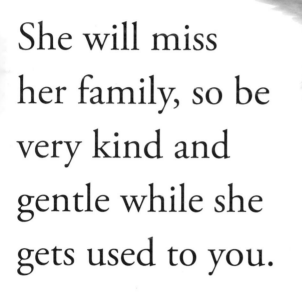

She will miss her family, so be very kind and gentle while she gets used to you.

Working dogs

Golden retrievers are intelligent dogs. They are quick learners and like to obey their owners, so they are easy to train.

This dog is helping to find a missing person.

Some work as search and rescue dogs. They help firemen and policemen find injured people after accidents.

Helping people

Golden retrievers are also trained as assistance dogs to help people with disabilities. They can learn to switch on lights, open doors and even help people to get dressed.

A golden retriever starred in the film *Homeward Bound: The Incredible Journey.*

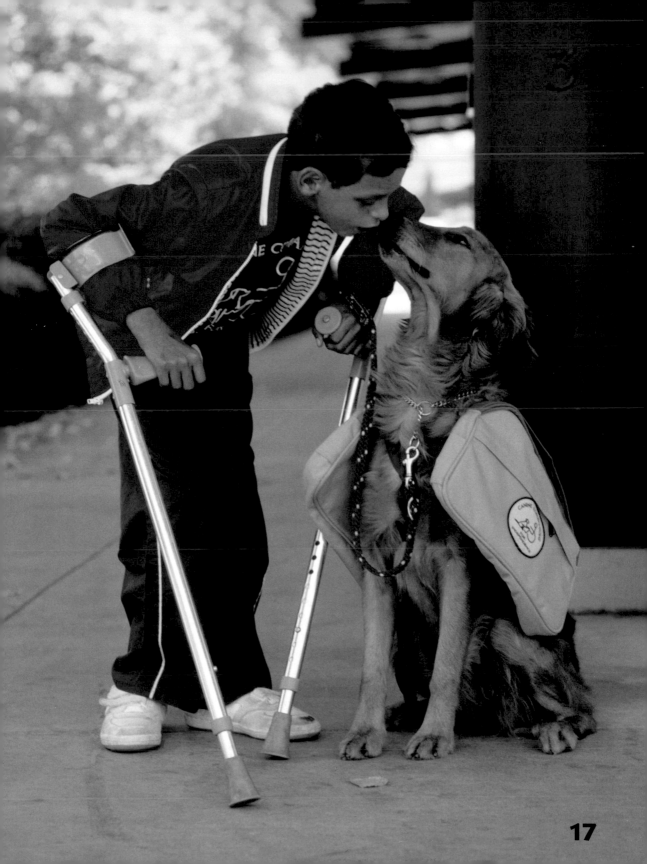

Water dogs

Golden retrievers are good swimmers and take any chance for a dip. They love to fetch sticks thrown into the water.

Never let your dog go for a swim on her own. She could run into trouble.

Your healthy golden retriever

These dogs can have hip problems, so you need to have a puppy checked before buying.

Your golden retriever will need brushing at least once a week to get rid of any loose hair. Check for fleas, too.

In summer, your dog may shed more hair and need brushing daily.

Golden retrievers will swim anywhere. Your dog might need a bath after a dip in a muddy pond.

Caring for your golden retriever

You and your family must think carefully before buying a golden retriever. She will live for at least ten years and need lots of attention.

Every day your dog will need food, water and exercise, as well as lots of love and care. You will also need to take her to the vet for regular checks and vaccinations. When you and your family go out or away on

holiday, you will have to make plans for your dog to be looked after.

Gerald Ford, who was president of the United States in the 1970s, owned a golden retriever named Liberty.

Useful words

muzzle
The long face of an animal such as a dog.

retrieve
To bring something back.

vaccination
An injection that protects your dog from certain illnesses.

Index

24